Drawing Circles

Use your compass to make some figures like these.

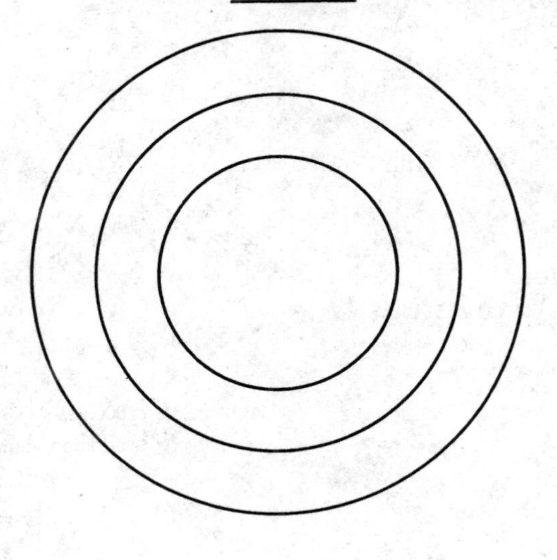

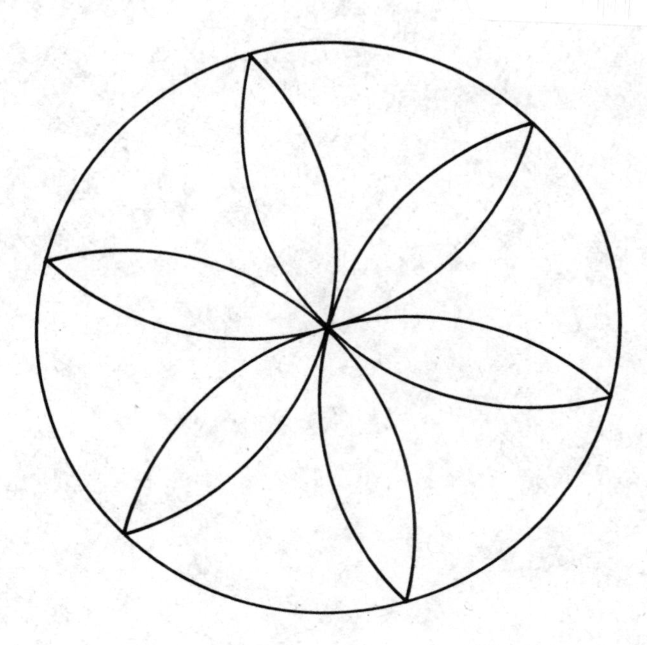

Use your compass and draw many circles of different sizes.

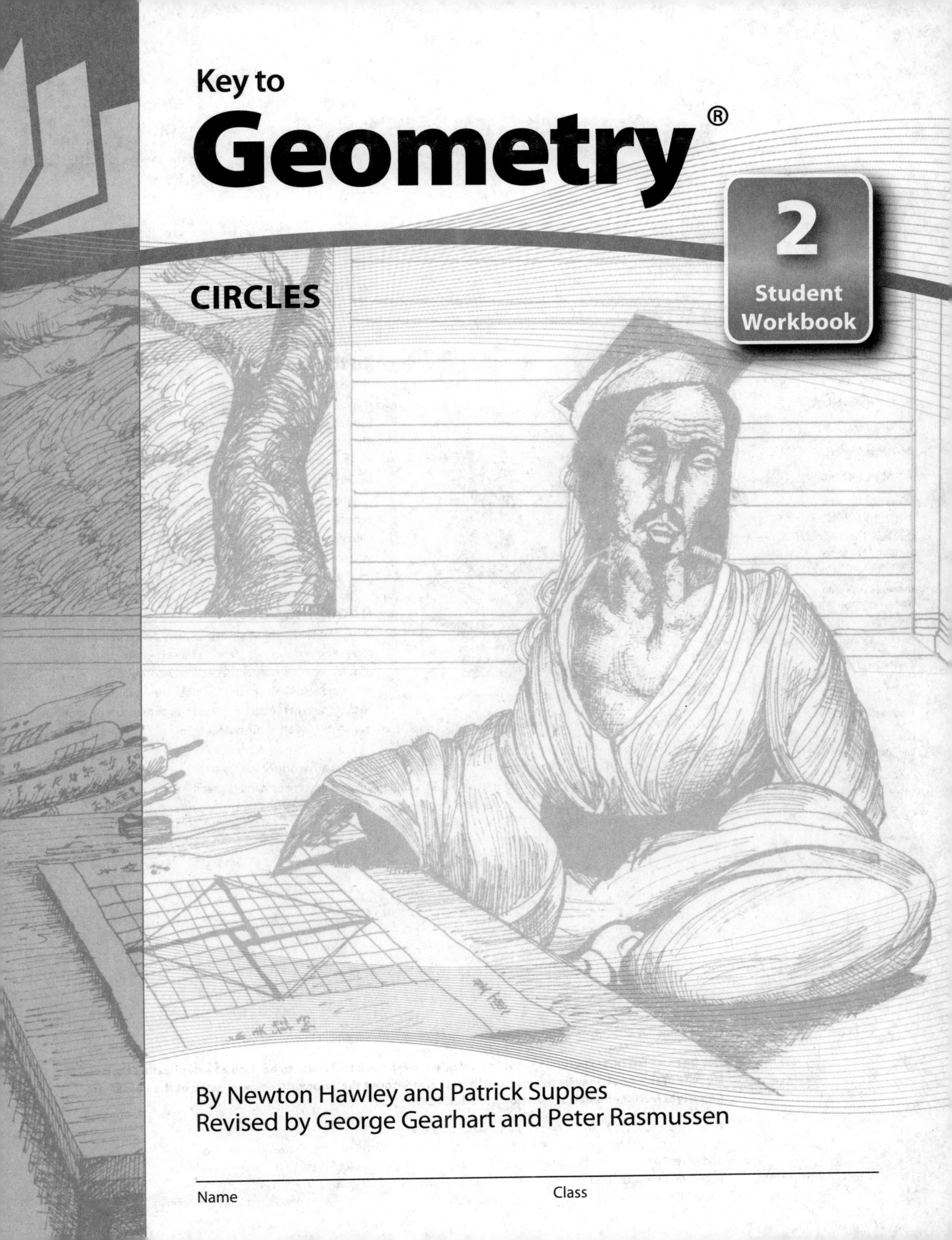

Key to

Geometry®

2 Student Workbook

CIRCLES

By Newton Hawley and Patrick Suppes
Revised by George Gearhart and Peter Rasmussen

Name Class

Key to

Geometry®

CIRCLES

TABLE OF CONTENTS

To the Student

These books will help you to discover for yourself many important relationships of geometry. Your tools will be the same as those used by the Greek mathematicians more than 2000 years ago. These tools are a *compass* and a *straightedge*. In addition, you will need a *sharpened pencil*. The lessons that follow will help you make drawings from which you may learn the most.

The answer books show *one* way the pages may be completed correctly. It is possible that your work is correct even though it is different. If your answer differs, re-read the instructions to make sure you followed them step by step. If you did, you are probably correct.

About the Cover

In many societies the study of geometry began as the art of measuring and surveying land. Chinese geometry began this way. Ancient Chinese geometers searched for ways to easily subdivide fields into rectangular plots. They found that rectangular plots could be created by combining right triangles. The *Chou Pei*, the oldest known Chinese mathematics text, begins with a discussion of the properties of right triangles. The exact age of the *Chou Pei* is not certain but it is at least 2000 years old.

On the cover of this booklet an ancient Chinese mathematician works on a page of the *Chou Pei*. The illustration represents a proof of an important theorem about right triangles. Although the theorem has been named after the Greek mathematician Pythagoras, Chinese geometers were familiar with the theorem and its proof centuries before the birth of Pythagoras.

Cover art by Howard Coale

Published by Key Curriculum Press, 1150 65th Street, Emeryville, CA 94608
ISBN 978-0-913684-72-6 12 LOV 22 21 Printed in the United States of America

1. Use your compass to draw a circle around the pentagon.

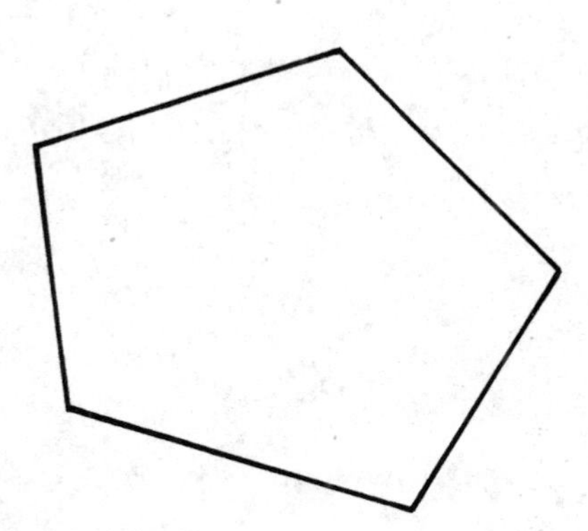

2. Use your compass to draw a circle around each quadrilateral: (only the quadrilaterals).

1. Draw several circles inside one another.

2. Draw a circle which has two of the figures inside it and has the third figure <u>outside</u>.

3. Underline the names of the figures inside the circle.

triangle quadrilateral pentagon

The Center of a Circle

1. Draw five different circles.

2. Draw a big circle, then draw a little circle inside it.

3. Draw a circle using point A as the center of the circle.

• A

1. Draw three circles with center B.

B •

2. Mark a point Z on the line segment. Draw a circle with point Z as its center.

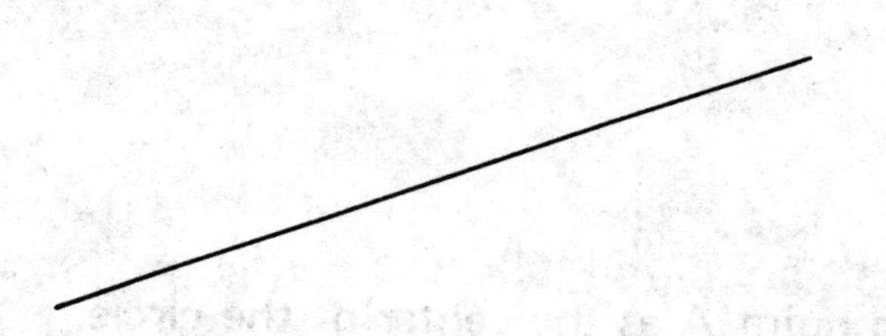

3. Mark a point S on the circle. Draw a circle which has S as its center.

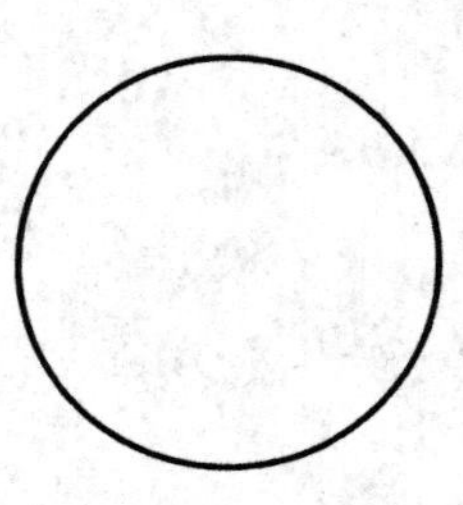

1. Draw a circle with center A which has the point B outside the circle.

A. . B . C

2. Now draw a circle with center C which has the point B outside it.

3. Draw a circle with center B which does not intersect either of the first two circles.

1. Draw a circle whose center is inside the pentagon.

2. Draw a circle with center A which has B inside the circle.

A. B. .C

Then draw a circle with center C which also has B inside the circle.

3. In how many points do these circles intersect? _ _ _ _ _

The Radius of a Circle

1. Draw a circle with center X.

• X

2. Draw a line segment from X to the circle.
   This line segment is a radius of the circle.

   A radius of a circle is a line segment from the center of the circle to the circle.

3. Draw a circle with center A.

A •

4. Draw a line segment from A to the circle.
5. Label as B the point where the radius meets the circle.

   $\overline{AB}$ is a _ _ _ _ _ _ _ _ _ _ _ _ _ _ _ of the circle.

6. Draw another radius of the circle.

1. Draw a circle with center C.

• C

2. Label as D a point on the circle.

3. Draw segment $\overline{CD}$.

   Is $\overline{CD}$ a radius of the circle? _ _ _ _ _ _ _

4. Draw another radius of the circle.

   How many radii can you draw? _ _ _ _ _ _ _ _ _ _ _ _ _ _ _ _ _

5. Draw two circles with center Z.

• Z

6. Draw a radius of the large circle.

   Label as X the point where it meets the large circle.

7. In how many points does radius $\overline{XZ}$ intersect the small circle? _ _ _ _ _

1. Draw a circle with center Q which passes through point P.

2. Draw a circle with center A which passes through point B.

3. Connect A and B.

   Is $\overline{AB}$ a radius of the circle? _ _ _ _ _ _

4. Label as C another point on the circle.

5. Draw $\overline{AC}$ and $\overline{BC}$.

   Is $\overline{AC}$ a radius of the circle? _ _ _ _ _ _

   Is $\overline{BC}$ a radius of the circle? _ _ _ _ _ _

Intersecting Circles

1. Draw a circle with center A.

• A

2. Draw a circle with center B which has point C inside the circle.

B • C •

3. Draw a circle with center C which has point B inside the circle.

4. Label as X and Y the points of intersection of the circles.

1. Draw a circle with center A which passes through point B.

A . . B

2. Draw a circle with center B which passes through A.

3. Label as X and Y the points of intersection of the two circles.

4. Draw the line segment determined by X and Y.

5. Draw the line segment determined by A and B.

1. Draw a circle with center A which passes through B.

A . . B

2. Draw a circle with center B which passes through A.

3. Label as P and Q the points of intersection.

4. Draw a circle with center P which passes through B.

5. Draw a circle with center Q which passes through A.

6. How many circles pass through point B? _ _ _ _ _ _

7. How many circles pass through point P? _ _ _ _ _ _

Comparing Segments

1. Draw three radii of the given circle.

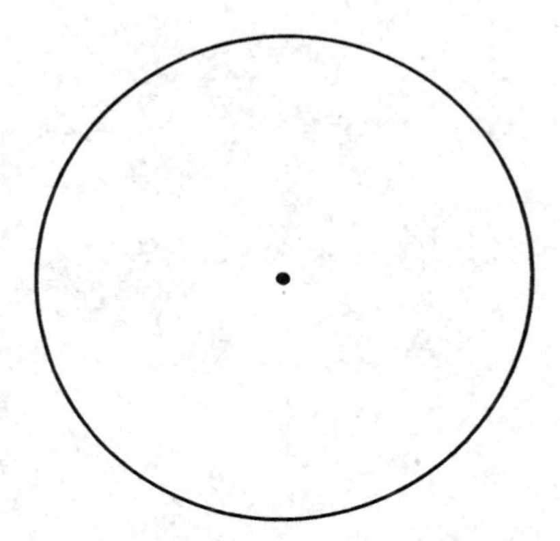

2. Are these radii all congruent? _ _ _ _ _

3. Draw line segment $\overline{CA}$.

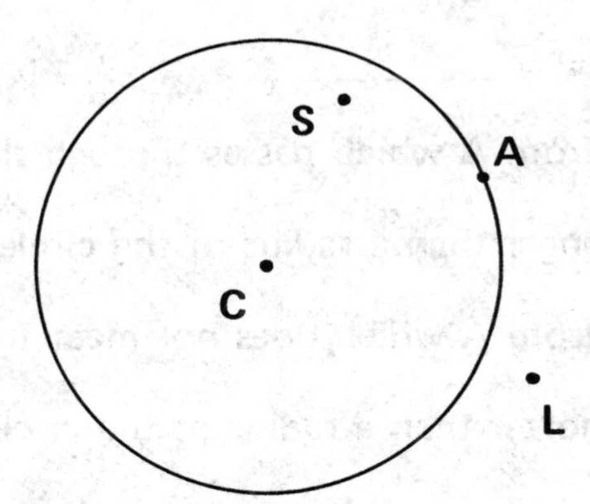

4. Is $\overline{CA}$ a radius of the circle? _ _ _ _ _

5. Connect C and L.

   Is $\overline{CL}$ <u>shorter</u> or <u>longer</u> than the radius $\overline{CA}$? _ _ _ _ _ _ _ _

   Is point L inside or outside the circle? _ _ _ _ _ _ _

6. Draw $\overline{CS}$.

   Is $\overline{CS}$ shorter or longer than the radius $\overline{CA}$? _ _ _ _ _ _ _

   Is point S inside or outside the circle? _ _ _ _ _ _ _

1. Draw a circle with center A.

A •

2. Draw two different radii of the circle.

   Which radius is longer? _

   Are these radii congruent? _ _ _ _ _ _

3. Draw a line segment from A which passes through the circle.

   Is this line segment longer than a radius of the circle? _ _ _ _ _

4. Draw a line segment from A which does not meet the circle.

   Is this line segment shorter than a radius of the circle? _ _ _ _ _

5. Which line segment is

   (a) <u>longest</u>? _ _ _ _ _

   (b) <u>shortest</u>? _ _ _ _ _

   (c) a radius? _ _ _ _ _

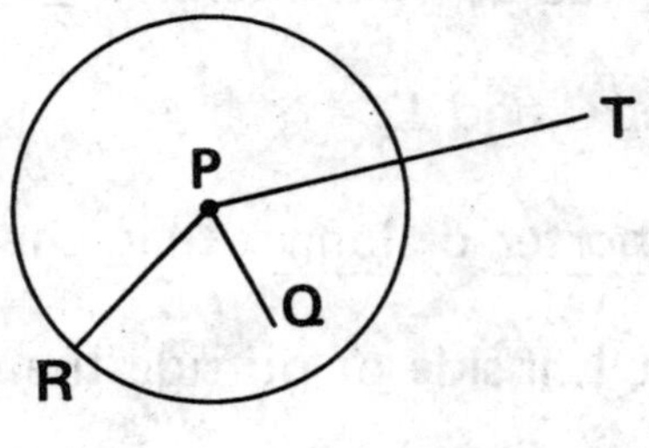

P T Q R

1. Draw a circle with center V.

• V

2. Draw two different radii of this circle.

3. Label as E and F the points where the radii meet the circle.

4. Connect E and F.

5. Draw a circle with center E and radius congruent to $\overline{EV}$.

6. Which is longer, $\overline{VE}$ or $\overline{EF}$? _ _ _ _ _

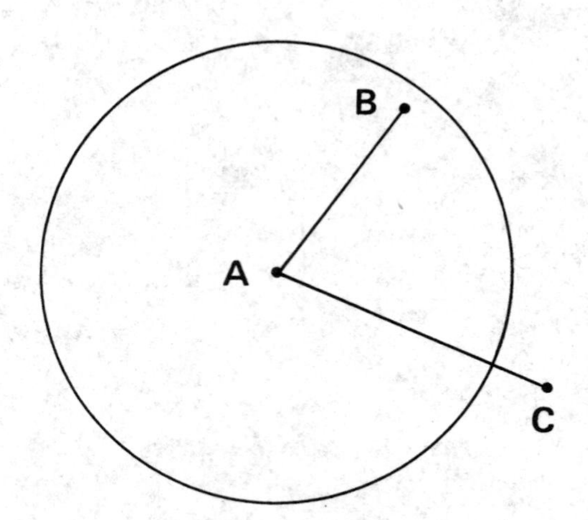

1. Are $\overline{AB}$ and $\overline{AC}$ congruent? _ _ _ _ _

2. Which one is longer? _ _ _ _ _

3. Draw a radius of the circle.

4. Is the radius longer or shorter than $\overline{AB}$? _ _ _ _ _ _ _ _

   Is the radius longer or shorter than $\overline{AC}$? _ _ _ _ _ _ _ _

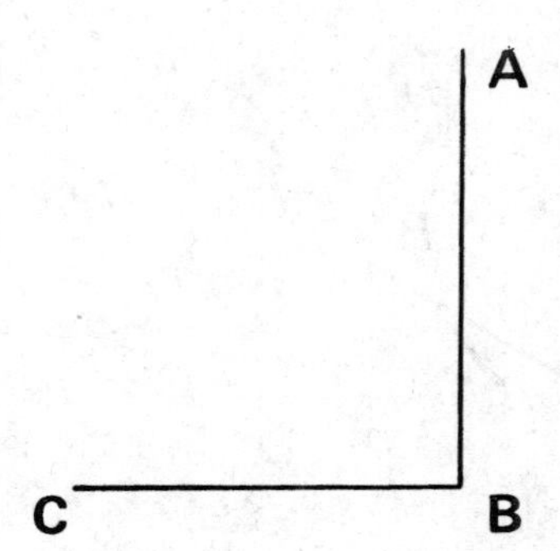

1. Draw a circle with center B which passes through C.

2. Which is longer, $\overline{BA}$ or $\overline{BC}$? _ _ _ _ _

3. Which is longer, $\overline{MN}$ or $\overline{MP}$? _ _ _ _ _
(Use your compass to decide.)

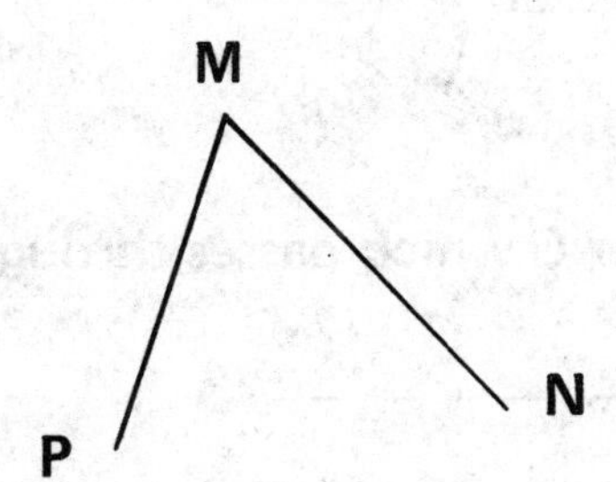

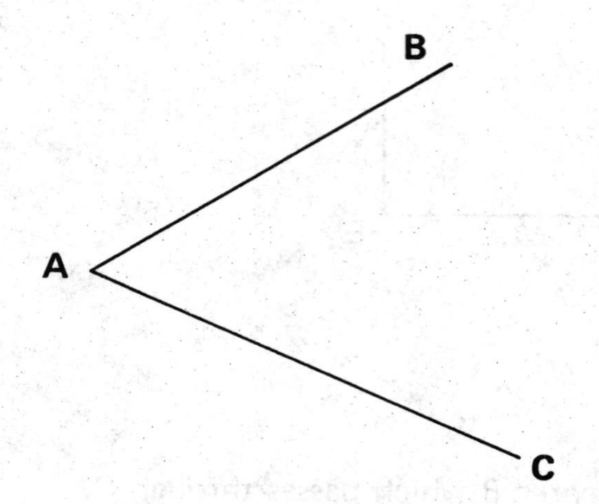

Use your compass to answer these questions.

1. Are $\overline{AB}$ and $\overline{AC}$ congruent? _ _ _ _ _

2. Which is longer? _ _ _ _ _

3. Draw $\overline{BC}$.

4. Draw a circle with center C which passes through B.

5. Is $\overline{BC}$ longer than $\overline{AC}$? _ _ _ _ _

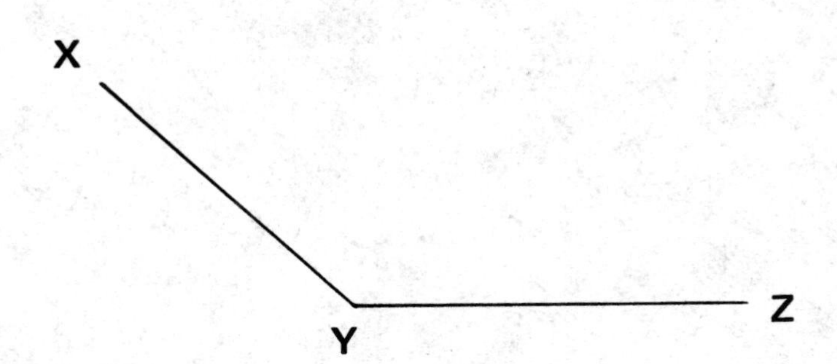

1. $\overline{XY}$ is _ $\overline{YZ}$.

   (a) longer than

   (b) congruent to

   (c) shorter than

2. Draw $\overline{XZ}$.

3. Now how many angles do you see? _ _ _ _ _

4. The figure XYZ is now a triangle.

   How many angles does it have? _ _ _ _ _

Circles with the Same Segment as Radius

1. Draw a circle with center A which passes through point B.

A. • B

2. Connect A and B.

3. $\overline{AB}$ is a _ _ _ _ _ _ _ _ _ _.

(a) spoke (c) rope

(b) spike (d) radius

4. Draw a circle with center B which passes through point A.

5. How many circles now have $\overline{AB}$ as a radius? _ _ _ _ _

1. Draw the segment determined by A and B.

A • • B

2. Draw a circle with center A which passes through B.

3. Is $\overline{AB}$ a radius of this circle? _ _ _ _ _

4. Draw another circle with $\overline{AB}$ as a radius.

5. Can you draw a third circle with $\overline{AB}$ as a radius? _ _ _ _ _

If your answer is Yes, then draw the circle.

Drawing a Segment Congruent to a Given Segment

1. Draw three circles with congruent radii.

2. Are these circles congruent? _ _ _ _ _

3. Draw a circle with center A and radius $\overline{AB}$.

C • ——— • D

A • ——— • B

4. Draw a circle with center C and radius congruent to $\overline{AB}$.

5. Does the second circle pass through point D? _ _ _ _ _

6. Are the circles congruent? _ _ _ _ _

7. Are $\overline{AB}$ and $\overline{CD}$ congruent? _ _ _ _ _ _

1. Draw a circle with center A and radius $\overline{AB}$.

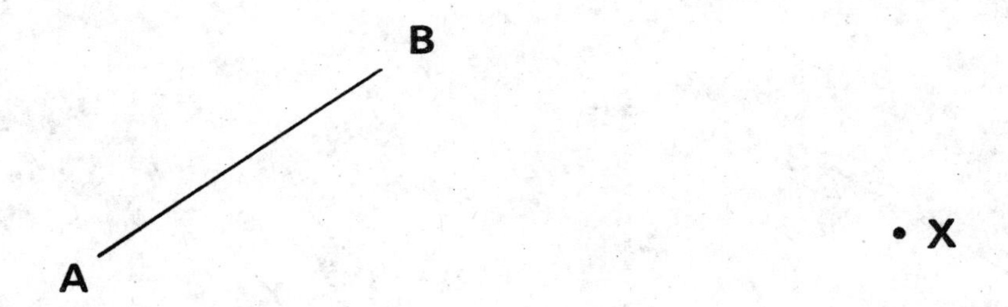

2. Now draw a circle with center X and radius congruent to $\overline{AB}$.

3. <u>Choose</u> a point on this last circle, and label the point Y.

4. Draw $\overline{XY}$.

5. $\overline{XY}$ is _ _ _ _ _ _ _ _ _ _ _ _ _ _ _ _ _ _ $\overline{AB}$.

   (a) shorter than

   (b) congruent to

   (c) longer than

6. The two circles are congruent because _ _ _ _ _ _ _ _ _ _ _ _ _

   _.

Problem: *Draw a segment using the straight edge and compass which is congruent to $\overline{AB}$ and which has X as an endpoint.*

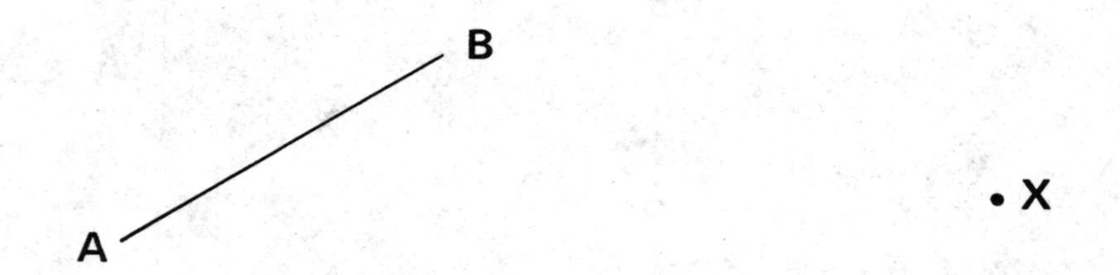

Solution:

1. Draw a circle with center X whose radius is congruent to $\overline{AB}$.
2. Draw a radius of this circle.

   Does this radius solve the problem? _ _ _ _ _

3. Draw a segment congruent to $\overline{EF}$ which has P as an endpoint.

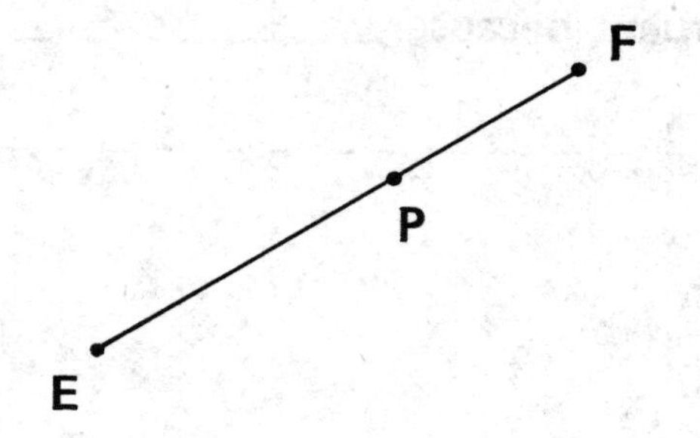

Drawing a Circle with Radius Congruent to a Given Segment

1. Draw a circle with $\overline{AB}$ as radius and B as center.

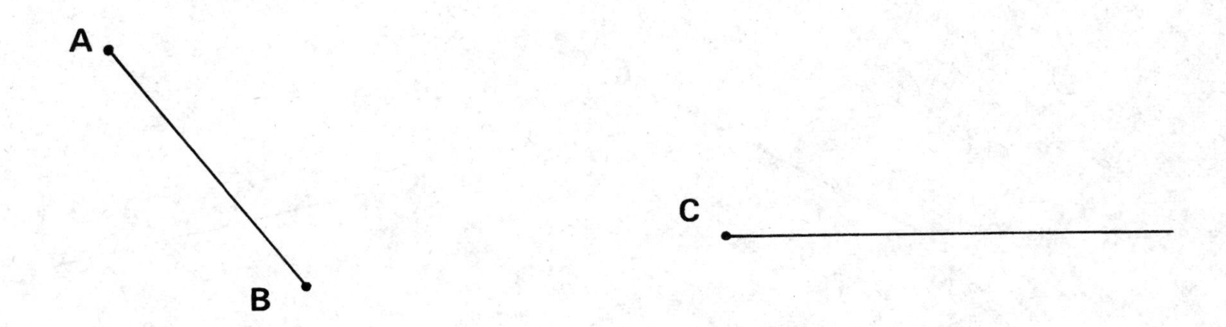

2. Draw a circle with center C and radius congruent to $\overline{AB}$.

3. Draw a circle with $\overline{PQ}$ as a radius.

4. Draw a circle with center R and radius congruent to $\overline{PQ}$.

   What can you say about $\overline{PQ}$ and $\overline{RS}$? _

1. Draw a circle with center K and radius congruent to $\overline{EF}$.

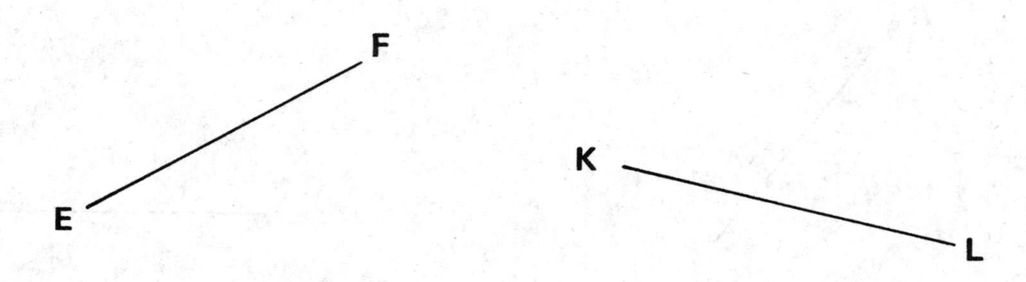

Which is longer, $\overline{EF}$ or $\overline{KL}$? _ _ _ _ _

2. Draw a circle with center C and radius congruent to $\overline{AB}$.

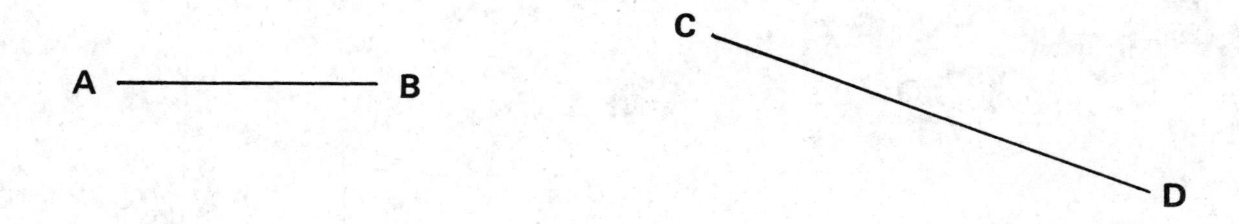

3. Label as E the point where this circle intersects $\overline{CD}$.

4. Is $\overline{CE}$ congruent to $\overline{AB}$? _ _ _ _ _

5. Is $\overline{CD}$ congruent to $\overline{AB}$? _ _ _ _ _

1. Draw a circle with center Z and radius $\overline{ZV}$.

   Is $\overline{ZY}$ congruent to $\overline{ZV}$? _ _ _ _ _

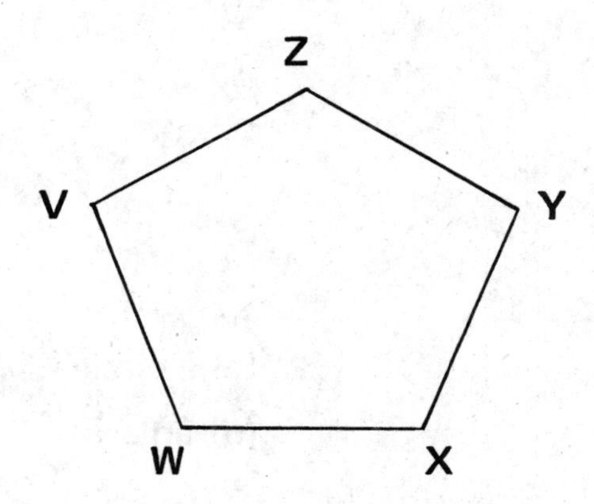

2. Draw a circle with center V and radius $\overline{ZV}$.

   Is $\overline{ZV}$ congruent to $\overline{VW}$? _ _ _ _ _

3. Draw a circle with center W and radius congruent to $\overline{ZV}$.

   Is $\overline{XW}$ congruent to $\overline{ZV}$? _ _ _ _ _

4. Draw a circle with center X and radius congruent to $\overline{ZV}$.

   Is $\overline{XY}$ congruent to $\overline{ZV}$? _ _ _ _ _

5. Draw a circle with center Y and radius congruent to $\overline{ZV}$.

6. The figure VWXYZ is a pentagon with all sides _ _ _ _ _ _ _ _ _ _

Arcs

An arc is a piece of a circle, just as a line segment is a piece of a line.

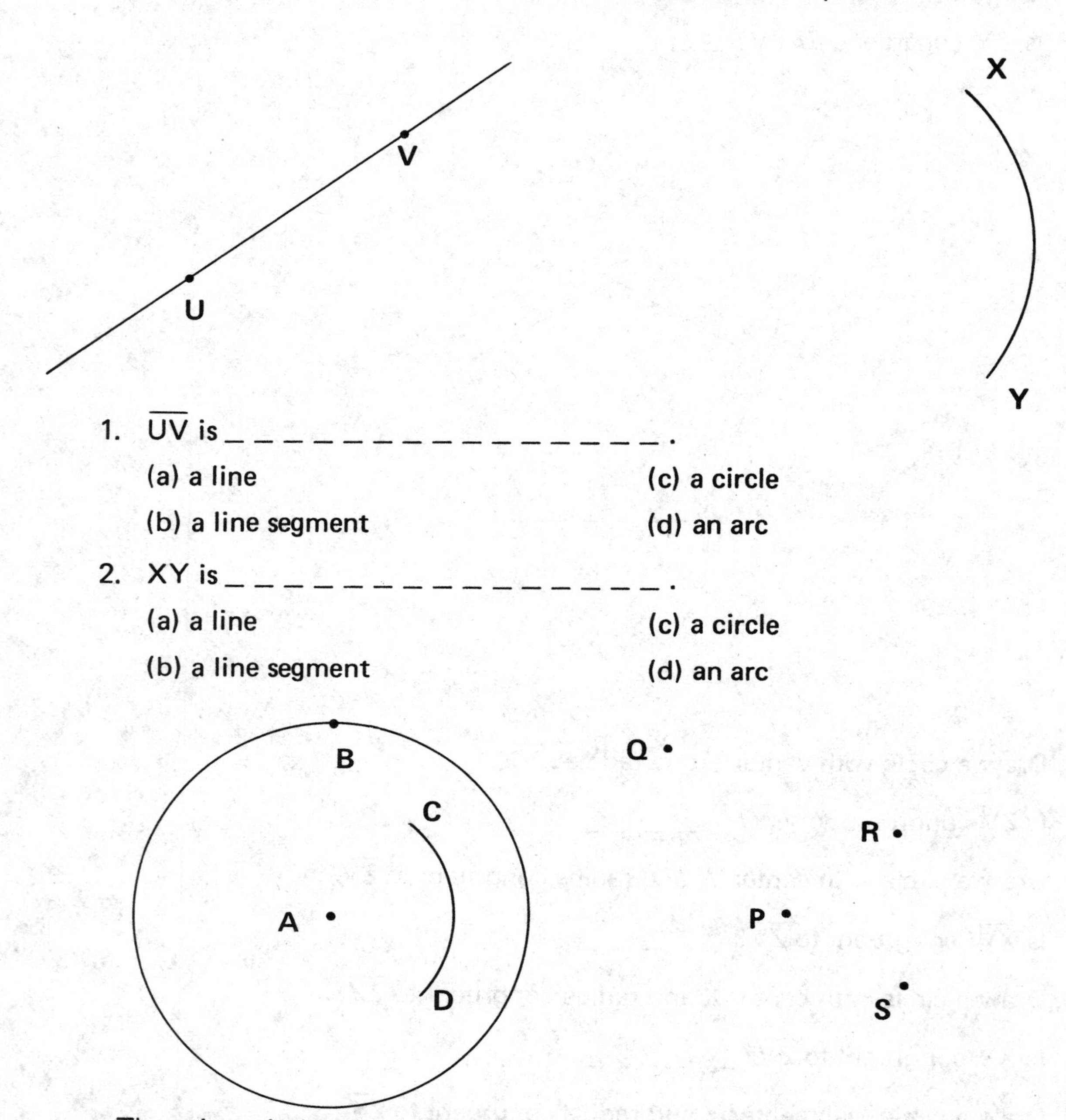

1. $\overline{UV}$ is __________________ .

   (a) a line (c) a circle

   (b) a line segment (d) an arc

2. XY is __________________ .

   (a) a line (c) a circle

   (b) a line segment (d) an arc

There is a circle with center A passing through B.
There is an arc drawn from C to D.
Its center is A because A is the center of the circle containing arc CD.

3. Draw a circle with center P passing through Q.
4. Draw an arc with center P from R to S.
5. Draw an arc with center C from A to B.
6. Draw the circle of which arc CD is a piece.

Here are two arcs with center A.

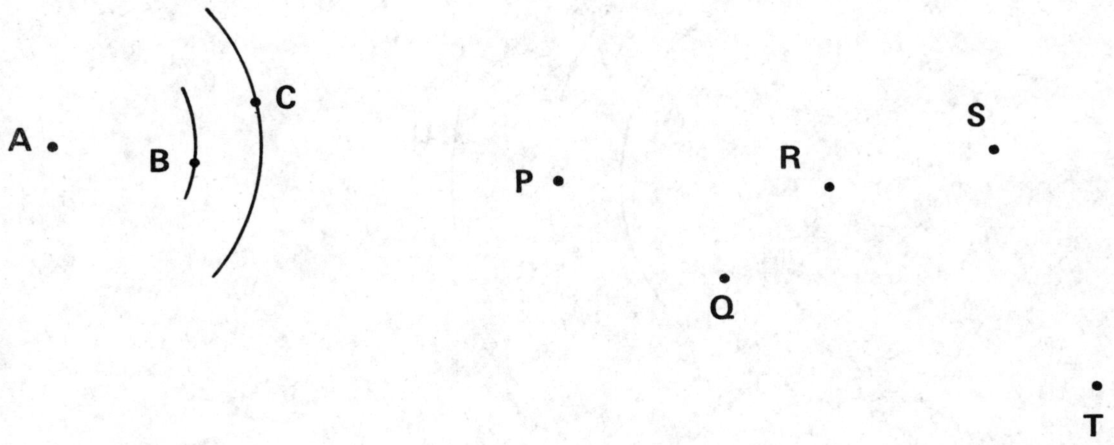

A • B C

P • R • S • Q • T •

There is a small arc passing through B.
There is a larger arc passing through C.

1. Draw a small arc with center P which passes through Q.
2. Draw a large arc with center P which passes through R.
3. Draw an arc with center P which passes through S.
4. Draw an arc with center P which passes through T.
5. Draw an arc with center Q which passes through R.
6. Draw the circle with center Q which passes through R.

The Radius of an Arc

The radius of an arc is the radius of the circle of which the arc is a piece.

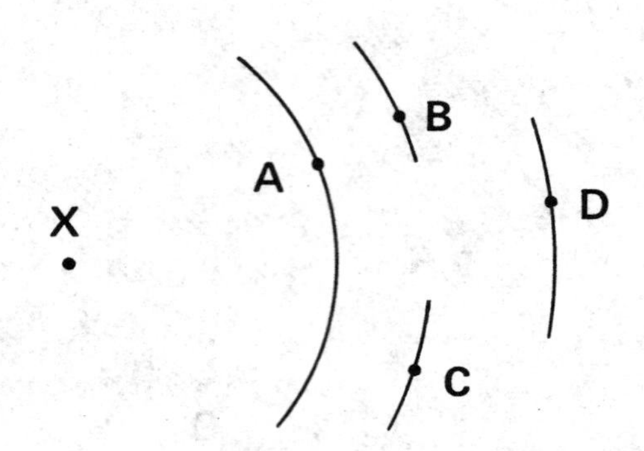

The large arc with center X passing through A has the smallest radius of the four arcs drawn.

1. Does the arc through C have the largest radius? _ _ _ _ _
2. Does the arc through B have a radius congruent to that of the arc through C? _ _ _ _ _
3. Which arc has the largest radius?

   The arc through _ _ _ _ .
4. Draw an arc with center X having a smaller radius than any of the other four.

1. Draw a radius of the arc.

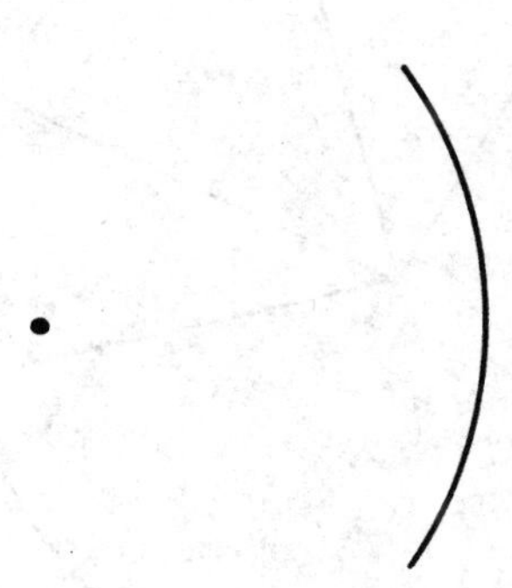

2. Draw an arc with center P and radius congruent to segment $\overline{RS}$.

3. Draw an arc with center A and radius congruent to $\overline{XY}$ which intersects $\overline{AB}$.

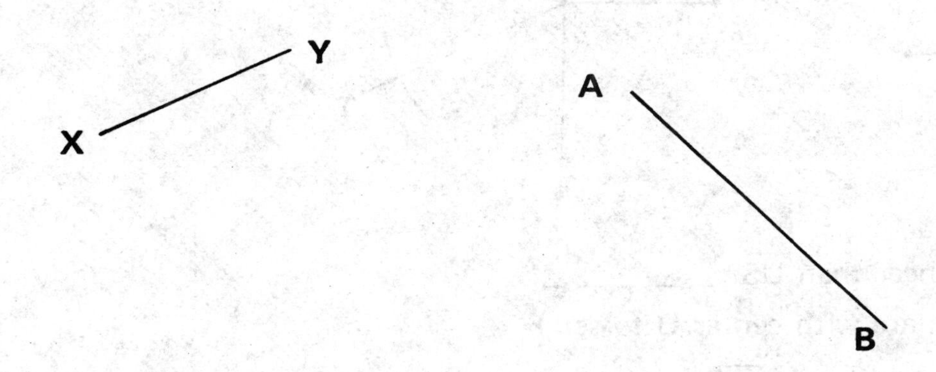

Comparing Segments

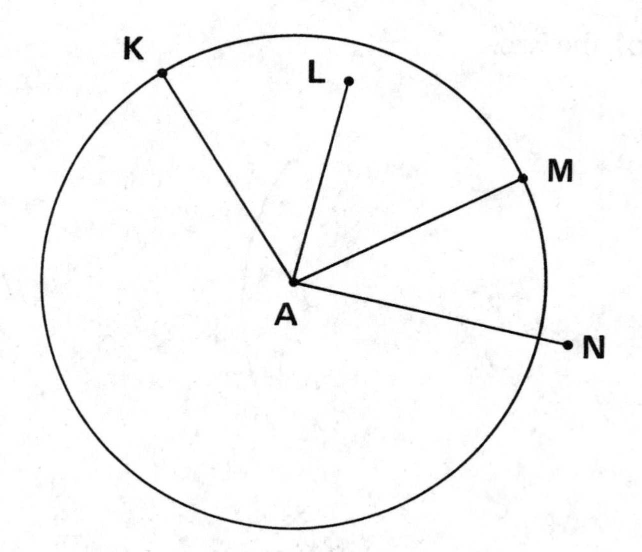

1. $\overline{AM}$ is congruent to ______ .

   $\overline{AM}$ is longer than ______ .

   $\overline{AM}$ is shorter than ______ .

2. The arcs with center U and radii congruent to $\overline{US}$ show us that:

   $\overline{US}$ is congruent to ______ .

   $\overline{US}$ is longer than ______ .

   $\overline{US}$ is shorter than ______ .

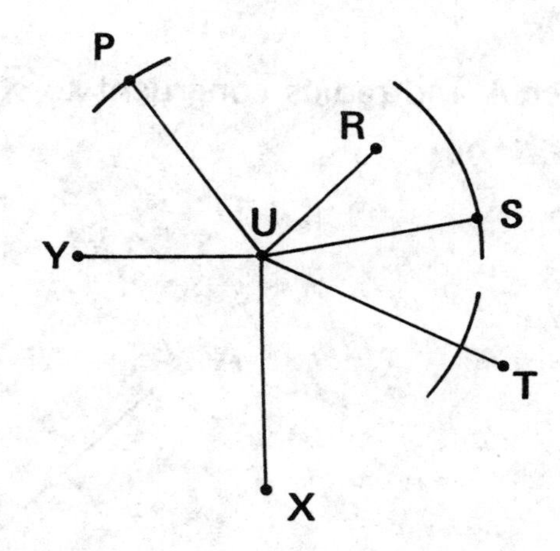

3. Is $\overline{UY}$ longer than $\overline{US}$? ______

   (Draw an arc with center U to see.)

4. Is $\overline{UX}$ congruent to $\overline{US}$? ______

1. Draw a small arc with center A and radius $\overline{AB}$ which intersects the ray through each segment.

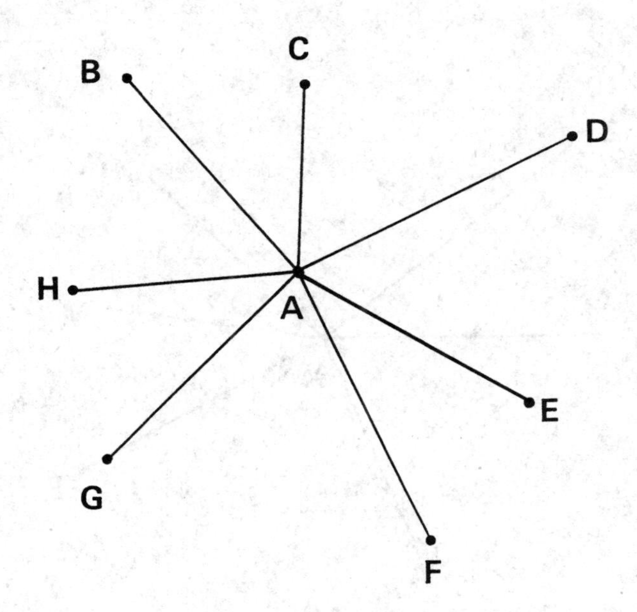

2. Which segments are congruent to $\overline{AB}$? _ _ _ _ _ _ _ _ _ _

   Which segments are shorter than $\overline{AB}$? _ _ _ _ _ _ _ _ _ _

   Which segments are longer than $\overline{AB}$? _ _ _ _ _ _ _ _ _ _

3. Draw an arc with center X and radius congruent to $\overline{PQ}$ which intersects $\overline{XY}$.

4. $\overline{PQ}$ is _ $\overline{XY}$.
   (a) shorter than
   (b) congruent to
   (c) longer than

1. Draw an arc with center X and radius $\overline{XY}$ which intersects the ray through each line segment.

2. Which line segment is congruent to $\overline{XY}$? _ _ _ _ _

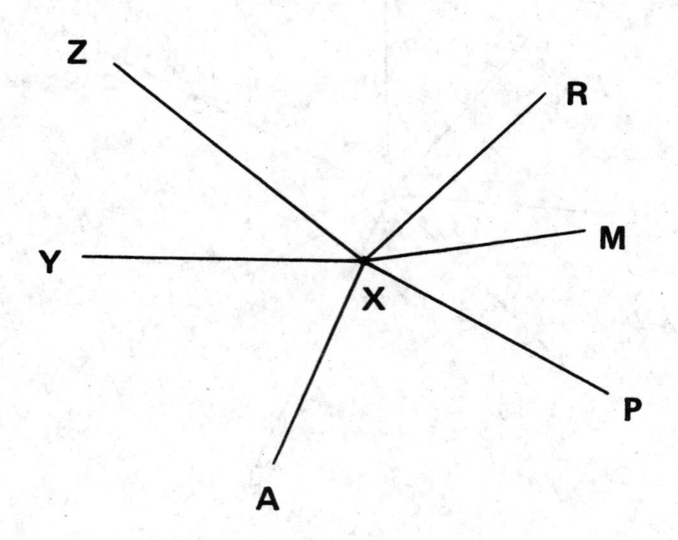

3. Are the two line segments congruent? _ _ _ _ _
(Draw arcs to see.)

4. The segment $\overline{UV}$ is _ _ _ _ _ _ _ _ _ _ _ _ _ _ _ _ _ $\overline{XY}$.
(a) longer than
(b) congruent to
(c) shorter than

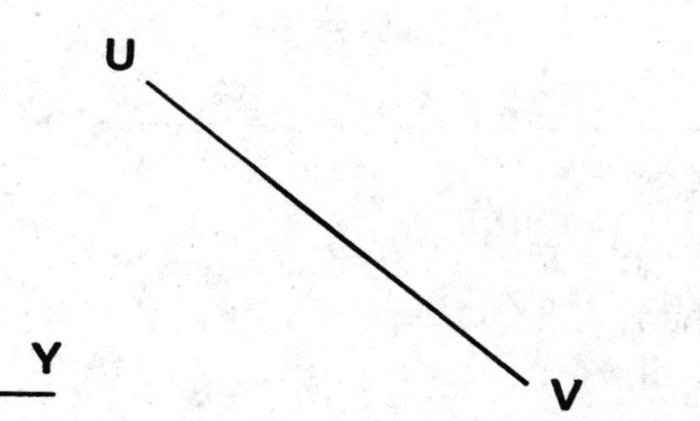

X Y

Equilateral Triangles

1. Are the three sides of this triangle congruent? _ _ _ _ _ _ _ _ _

   (Draw arcs to see.)

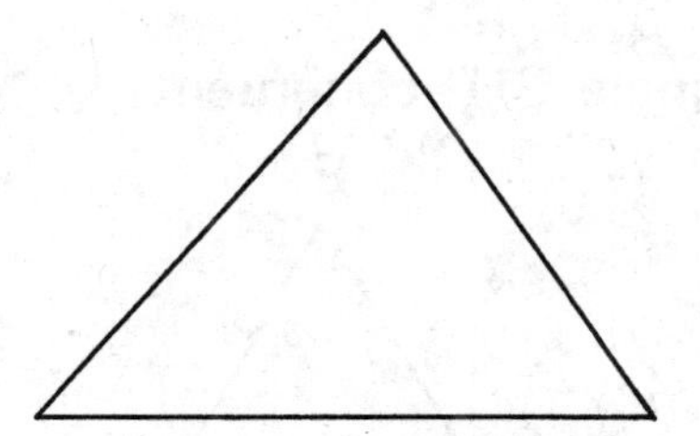

2. Draw an arc with center A and radius $\overline{AB}$ intersecting $\overline{AB}$.

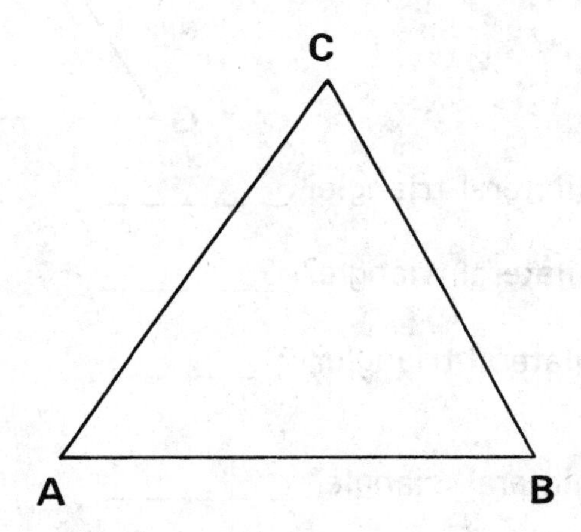

3. Draw an arc with center A and radius $\overline{AB}$ intersecting ray $\overrightarrow{AC}$.

   Does this arc pass through C? _ _ _ _ _

   Is segment $\overline{AB}$ congruent to segment $\overline{AC}$? _ _ _ _ _

4. Draw an arc with center C and radius congruent to $\overline{AB}$ intersecting ray $\overrightarrow{CB}$.

   Is $\overline{AB}$ congruent to $\overline{BC}$? _ _ _ _ _

5. What can you say about the three segments in triangle ACB? _ _ _

   _

A triangle is equilateral if all its sides are congruent.

1. Are all three sides of triangle ABC congruent? _ _ _ _ _

2. Are all three sides of triangle DEF congruent? _ _ _ _ _

3. Are all three sides of triangle GHI congruent? _ _ _ _ _

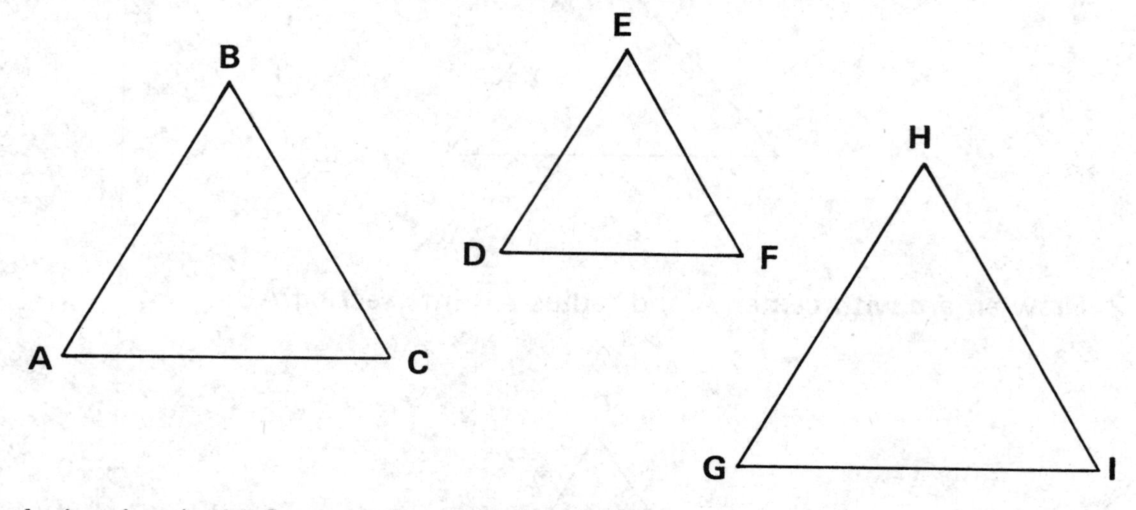

4. Is triangle ABC an equilateral triangle? _ _ _ _ _

5. Is triangle DEF an equilateral triangle? _ _ _ _ _

6. Is triangle GHI an equilateral triangle? _ _ _ _ _

7. Is triangle RST an equilateral triangle? _ _ _ _ _

8. Is triangle UVW an equilateral triangle? _ _ _ _ _

9. Is triangle XYZ an equilateral triangle? _ _ _ _ _ _
(Do not guess.)

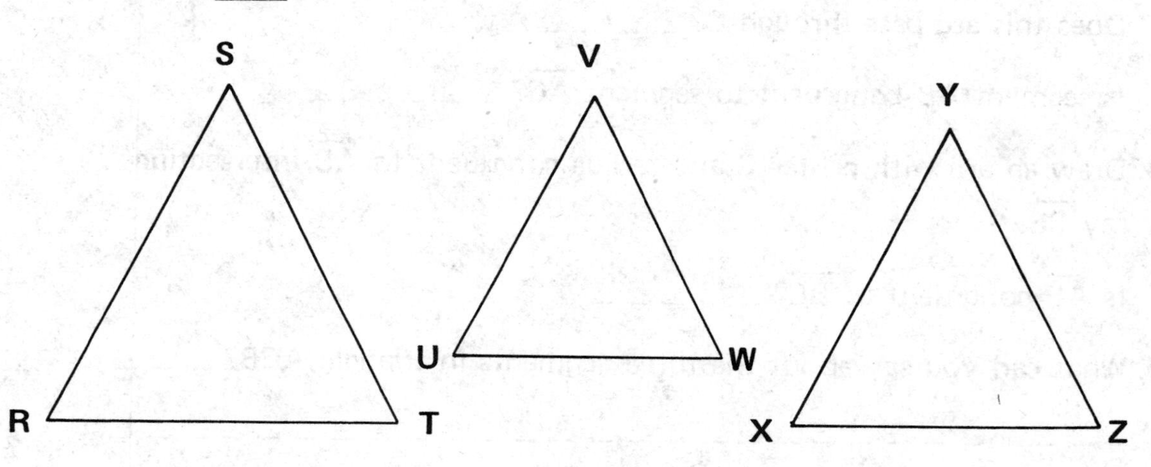

1. Which line segment is congruent to $\overline{XY}$? _ _ _ _ _

(Draw arcs with radius congruent to $\overline{XY}$ to find out.)

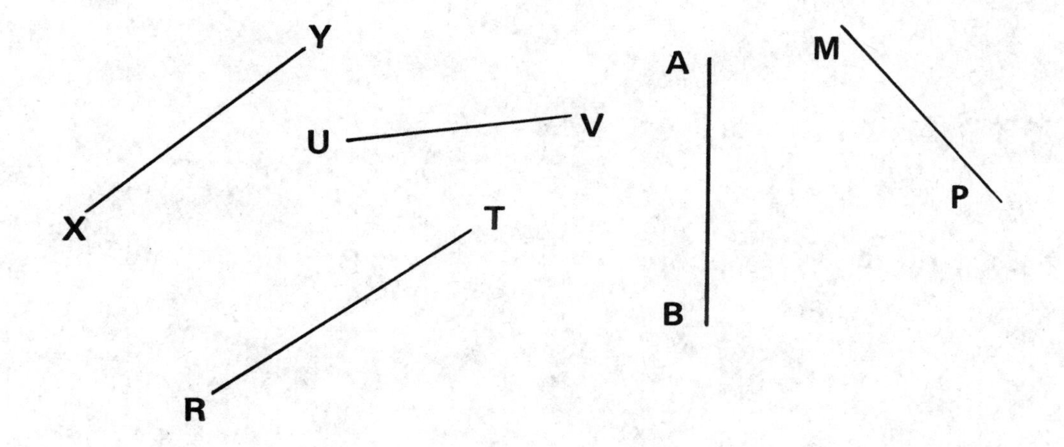

2. Which triangle is equilateral? _ _ _ _ _

(Draw arcs to find out.)

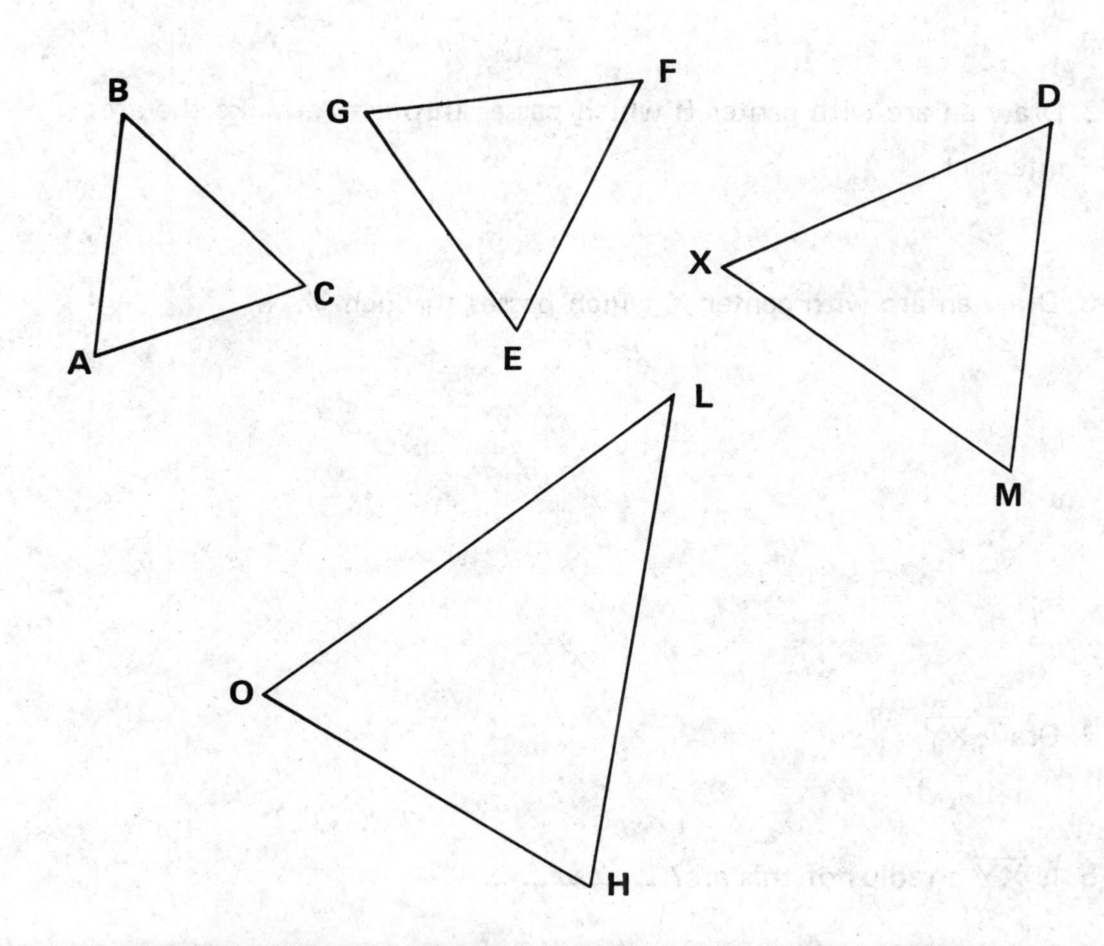

1. Draw a large arc with center A which passes through B.

A . . B

2. Draw an arc with center B which passes through A. Make the arcs intersect.

3. Draw an arc with center X which passes through Y.

. Y

X .

4. Draw $\overline{XY}$.

5. Is $\overline{XY}$ a radius of this arc? _ _ _ _ _

• B

A •

1. Draw an arc above B with center A and radius $\overline{AB}$.
2. Draw another arc with center B and radius $\overline{AB}$.
   Make the arcs intersect.
3. Label the point of intersection C.
4. Draw triangle ABC.
5. What kind of triangle is ABC? _ _ _ _ _ _ _ _ _ _ _ _ _ _ _ _ _ _

   How do you know? _

Intersecting Circles and Triangles

1. Draw a circle with center A which passes through B.

A •   • B

2. Draw the segment determined by A and B.

3. Is $\overline{AB}$ a radius of the circle? _ _ _ _ _

4. Now draw a circle with center B which passes through A.

5. Do the two circles intersect? _ _ _ _ _

6. Label as C and D the points of intersection.

7. Draw $\overline{AC}$ and $\overline{AD}$.

8. Is $\overline{AC}$ congruent to $\overline{AB}$? _ _ _ _ _

9. Is $\overline{AD}$ longer than $\overline{AB}$? _ _ _ _ _

10. How do you know that $\overline{AC}$, $\overline{AB}$, and $\overline{AD}$ are all congruent? _ _ _ _

_ _

1. Draw a circle with center A which passes through B.

A B

2. Draw a circle with center B which passes through A.

3. Label as C and D the points of intersection of the circles.

4. Draw $\overline{CD}$.

5. $\overline{CD}$ is _ _ _ _ _ _ _ _ _ _ _ _ _ _ _ _ _ _ _ $\overline{AB}$.

   (a) shorter than
   (b) congruent to
   (c) longer than

6. Draw two circles which have $\overline{EF}$ as a radius.

E F

1. Draw two circles which have $\overline{PQ}$ as a radius.

Q

P

2. Label as S and T the points of intersection of the two circles.

3. Draw $\overline{PQ}$ and $\overline{QT}$ and $\overline{PT}$.

4. Is the triangle TPQ equilateral? _ _ _ _ _

5. Draw the segment determined by S and T.

6. Draw an arc with center S and radius congruent to $\overline{PQ}$ which intersects $\overline{ST}$.

7. $\overline{ST}$ is _ _ _ _ _ _ _ _ _ _ _ _ _ _ _ _ _ _ _ $\overline{PQ}$.

   (a) shorter than
   (b) longer than
   (c) congruent to

Constructing an Equilateral Triangle

1. Draw an arc with center A and radius $\overline{AB}$.

A —————————————— B

2. Now draw an arc with center B and radius $\overline{AB}$.
   Make the two arcs intersect above $\overline{AB}$.

3. Label as C this point of intersection.

4. Draw $\overline{AC}$. Then draw $\overline{BC}$.

5. Is $\overline{AB}$ congruent to $\overline{AC}$? _ _ _ _ _

6. Is $\overline{AB}$ congruent to $\overline{BC}$? _ _ _ _ _

7. Is $\overline{AC}$ congruent to $\overline{BC}$? _ _ _ _ _

8. What do we call triangle ABC? _ _ _ _ _ _ _ _ _ _ _ _ _ _ _ _ _

1. Draw the circles which have $\overline{AB}$ as a radius.

2. In how many points do these circles intersect? _ _ _ _ _

3. Label these points C and D.

4. Draw $\overline{AC}$. Then draw $\overline{BC}$.

5. Is $\overline{AB}$ congruent to $\overline{AC}$? _ _ _ _ _

   Is $\overline{AB}$ congruent to $\overline{BC}$? _ _ _ _ _

   Is $\overline{AC}$ congruent to $\overline{BC}$? _ _ _ _ _

6. What kind of triangle is ABC? _ _ _ _ _ _ _ _ _ _ _ _ _ _ _ _

1. Draw the circles which have $\overline{EF}$ as a radius.

E

F

2. Label the points of intersection G and H.

3. Draw all segments determined by two of the points E, F, G, and H.

4. How many triangles can you find in this figure? _ _ _ _ _ _

5. How many equilateral triangles can you find? _ _ _ _ _ _

6. Name the equilateral triangles. _

Constructing Hexagons

1. Draw the circle with center A and radius $\overline{ZA}$.

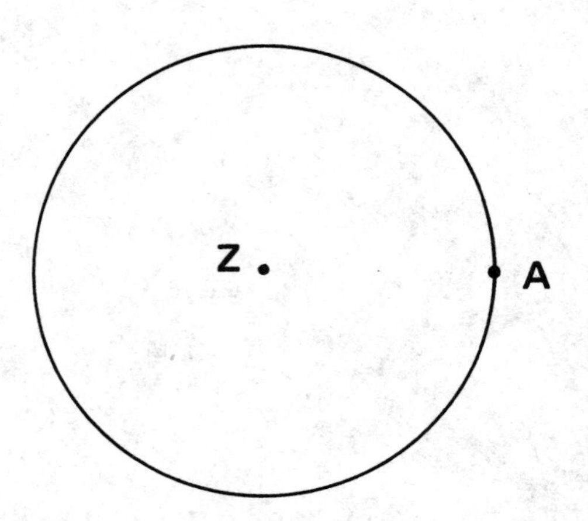

2. Label the points of intersection B and F.
3. Draw the circle with center B which passes through Z.
4. Label as C its new point of intersection with the circle of center Z.
5. Draw a circle with center C which passes through Z.
6. Label as D its new point of intersection with the circle of center Z.
7. Draw a circle with center D which passes through Z.
8. Label the new point of intersection E.
9. Draw a circle with center E which passes through Z.
10. Draw a circle with center F which passes through Z.
11. Draw $\overline{AB}$, $\overline{BC}$, $\overline{CD}$, $\overline{DE}$, $\overline{EF}$, $\overline{FA}$.
12. What do we call figure ABCDEF? _ _ _ _ _ _ _ _ _ _ _ _ _ _ _ _

1. Draw a circle with center P.

P

2. Draw a line through P.
3. Label as R and S the points of intersection.
4. Draw a circle with center R and radius $\overline{PR}$.
5. Label as T and W the points of intersection.
6. Draw a circle with center S and radius $\overline{PS}$.
7. Label as X and Y the points of intersection.
8. Draw the hexagon through R, W, X, S, Y, and T.
9. Draw triangles PXS, PYS, PTR, and PRW.
10. What kind of triangles are they? ____________________

Review

• B

A •

• D

• C

1. Draw an arc with center A from B to C.

2. Draw an arc with center A passing through D.

   What conclusion can you make regarding the two circles formed by the arcs? _

3. Are the segments $\overline{XY}$ and $\overline{XZ}$ radii of the same circle? _ _ _ _ _

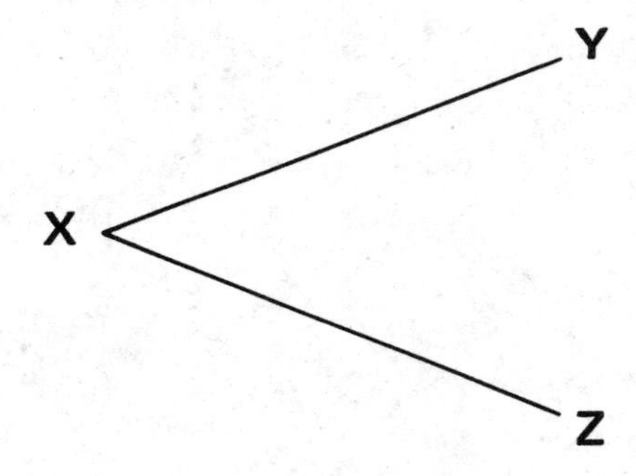

4. Compare the line segments $\overline{RS}$ and $\overline{UV}$.

   Are they congruent? _ _ _ _ _

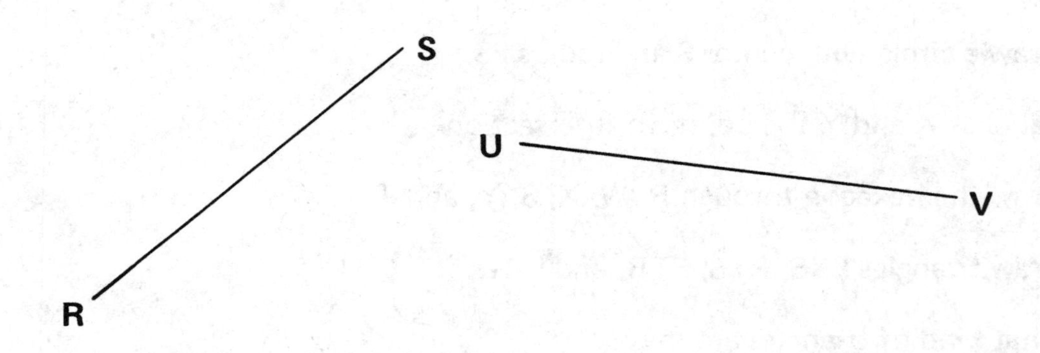

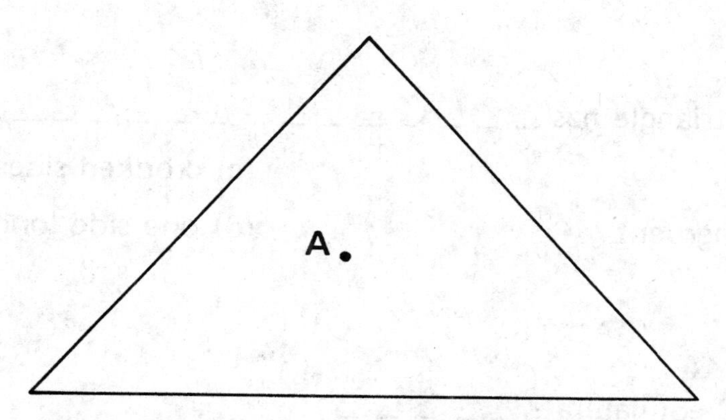

1. Draw a circle which is inside the triangle and has center A.

2. Draw a circle with center B which has the quadrilateral inside it.

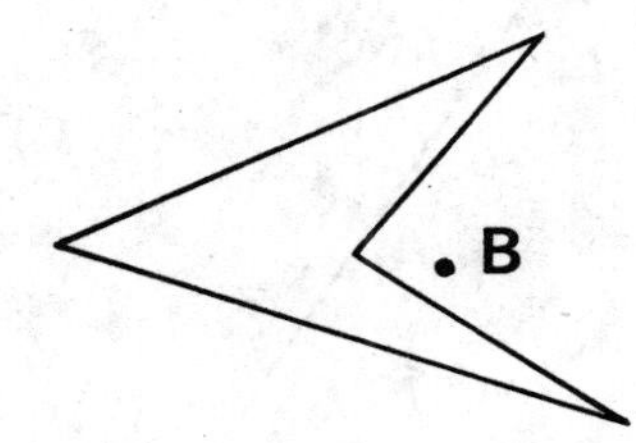

3. Draw a circle having the given line segment as a radius.

4. Draw a circle with center X passing through Y.

5. What is segment $\overline{XY}$ called? _ _ _ _ _ _ _ _ _ _ _ _

Practice Test

1. An equilateral triangle has _ _ _ _ _ _ _ _ _ _ _ _ _ _ _ _

   (a) four sides
   (b) all sides congruent
   (c) crooked sides
   (d) one side longer

2. Is triangle ABC equilateral? _ _ _ _ _

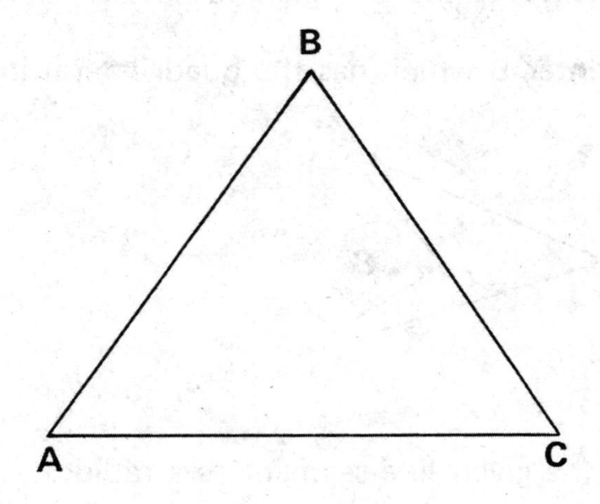

3. The radii of a circle are all _ _ _ _ _ _ _ _ _ _ _ _ _ _ _ _ _ .

   (a) centers
   (b) congruent
   (c) arcs

4. The segment $\overline{EF}$ is _ _ _ _ _ _ _ _ _ _ _ _ _ _ _ _ $\overline{EG}$.

   (a) longer than
   (b) congruent to
   (c) shorter than

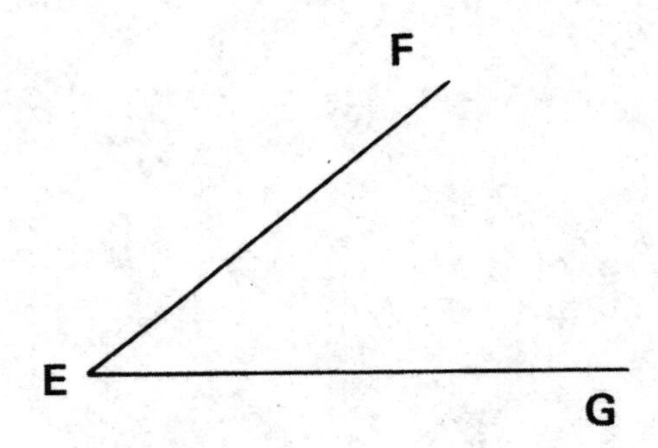

5. Are the two segments congruent? _ _ _ _ _ _

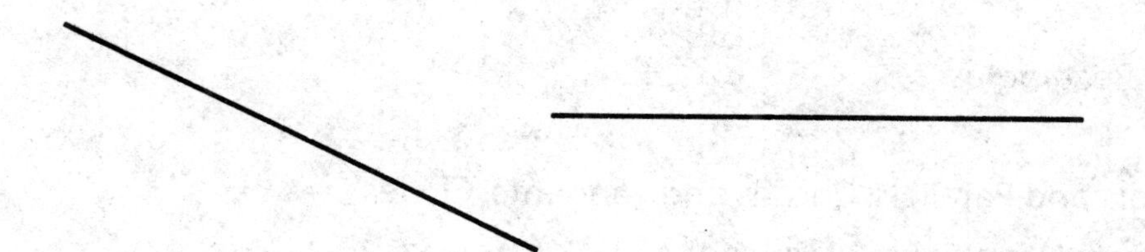

6. Draw the circles which have $\overline{PQ}$ as a radius.

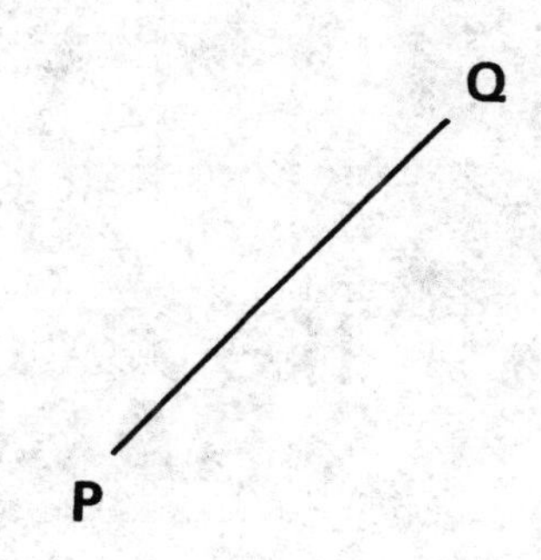

7. Label one point of intersection R.

8. Draw $\overline{PR}$ and $\overline{QR}$.

9. Is $\overline{PQ}$ congruent to $\overline{PR}$? _ _ _ _ _ _

   Is $\overline{PQ}$ congruent to $\overline{QR}$? _ _ _ _ _ _

   Is $\overline{PR}$ congruent to $\overline{QR}$? _ _ _ _ _ _

10. What kind of triangle is PQR? _ _ _ _ _ _ _ _ _ _ _ _ _ _ _ _

Key to Geometry® workbooks

Book 1: Lines and Segments
Book 2: Circles
Book 3: Constructions
Book 4: Perpendiculars
Book 5: Squares and Rectangles
Book 6: Angles
Book 7: Perpendiculars and Parallels, Chords and Tangents, Circles
Book 8: Triangles, Parallel Lines, Similar Polygons

Answers and Notes for Books 1–3
Answers and Notes for Books 4–6
Answers and Notes for Book 7
Answers and Notes for Book 8

The KeyTo TRACKER™

The Online Companion to the *Key to...*® Workbook Series

Save time, improve learning, and monitor student progress with The Key To Tracker, the online companion to the Key to... workbooks for fractions, decimals, percents, and algebra.

Learn more: www.keypress.com/keyto

Also available in the Key to...® series

Key to Fractions®
Key to Decimals®
Key to Percents®
Key to Algebra®
Key to Metric®
Key to Metric Measurement®

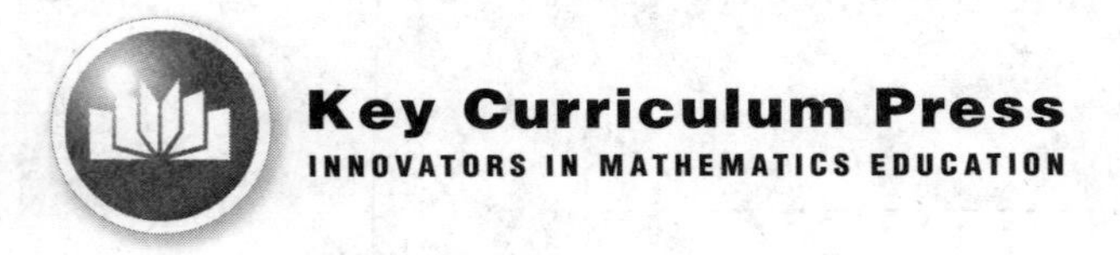

ISBN 978-0-913684-72-6